YOUR KNOWLEDGE HAS VALUE

Donal Yeang

Energy Conservation in Cambodia and ASEAN

GRIN Verlag

Bibliografische Information der Deutschen Nationalbibliothek:

Die Deutsche Bibliothek verzeichnet diese Publikation in der Deutschen National-
bibliografie; detaillierte bibliografische Daten sind im Internet über http://dnb.d-
nb.de/ abrufbar.

Imprint:

Copyright © 2011 GRIN Verlag GmbH
Druck und Bindung: Books on Demand GmbH, Norderstedt Germany
ISBN: 978-3-640-90171-5

This book at GRIN:

http://www.grin.com/en/e-book/170883/energy-conservation-in-cambodia-and-
asean

GRIN - Your knowledge has value

Der GRIN Verlag publiziert seit 1998 wissenschaftliche Arbeiten von Studenten, Hochschullehrern und anderen Akademikern als eBook und gedrucktes Buch. Die Verlagswebsite www.grin.com ist die ideale Plattform zur Veröffentlichung von Hausarbeiten, Abschlussarbeiten, wissenschaftlichen Aufsätzen, Dissertationen und Fachbüchern.

Visit us on the internet:

http://www.grin.com/

http://www.facebook.com/grincom

http://www.twitter.com/grin_com

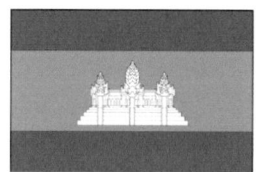

Country Report

"The Third Training Program on Energy Conservation to Reduce Global Warming for ASEAN Countries"

14-25 February 2011

Bangkok, Thailand

Energy Conservation in Cambodia and ASEAN

Prepared by

Mr. Donal Yeang

Organized by

Thailand International Development Cooperation Agency

สำนักงานความร่วมมือเพื่อการพัฒนาระหว่างประเทศ

Ministry of Foreign Affairs

กระทรวงการต่างประเทศ

And

Energy Conservation in Cambodia and ASEAN

1. Geography

Situated in the tropical region of Southeast Asia, the kingdom of Cambodia covers an area of 181 035 square kilometers (69 898 sq mi), with 800 Km border with Thailand in the west, 450 Km with Lao PDR in the north, 1 250 Km with Viet Nam in the east and a coastline of 440 Km long. The physical landscape is dominated by the lowland plains around the Mekong River and the Tonle Sap Lake. Approximately 49% remains covered by forest and there are about 2.5 million hectares of arable land and over 0.5 million hectares of pasture land. Cambodia's climate is tropical and subject to both southeast and northwest monsoons. The southeast monsoon, which coincides with the rainy season, extends from May to October. The northwest monsoon brings a cool but drier period from November to April. The average annual rainfall is about 1500 mm, with the heaviest rainfalls of up to 4000 mm per year occurring in the southwest coastal line. The temperatures are fairly uniform in the central basin area with an average of about 27°C. The maximum temperatures in the region vary from 35°C to 38°C which are common before the start of the rainy season, but the temperatures very rarely fall below 10°C.

Map of Cambodia

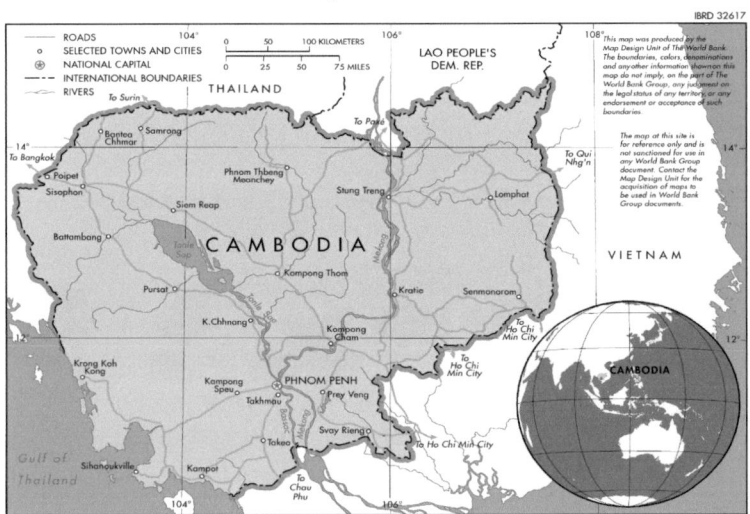

Source: Worldbank, 2003

2. Socio-economic Situation

The population of Cambodia is estimated about 14.6 million people of which over 80 percent lives in the rural areas (World Bank, 2010). The population ethnically consists of about 90% Khmer, 5% of Vietnamese, 1% of Chinese, and 4% other (CIA, 2010). The rural populations are engaged in agriculture, fishing, forest harvesting for their livelihoods. The GDP recorded a growth of 4.5% in the year 2000 (lower than the 5.5% projected, but higher than the 4% in 1999, 1.8% in 1998, and 3.7% in 1997), and in 2001 and 2002, a growth respectively of 6% and 5.5%. From 2004 to 2007, the economy grew about 10% per year, driven largely by an expansion in the garment sector, construction, agriculture, and tourism. Due to the global economic slowdown, GDP lowed to recorded 1.5% in 2009, but climbed more than 4% in 2010, driven by renewed exports. With the January 2005 expiration of a WTO Agreement on Textiles and Clothing, Cambodian textile producers were forced to compete directly with lower-priced countries such as China, India, Vietnam, and Bangladesh. In addition, oil and mineral deposits were found in Cambodia in 2005 and the government is hoping to generate new revenue from these extractive industries in the near future. The government has reported that bauxite, gold, iron and gems deposits are discovered in the country. As a result, mining sector is becoming the most attractive business for both domestic and foreign investors. In 2009, rubber exports increased about 25% due to rising global demand and continued to increase up to 35.4% in 2010 (Cambodia Daily, 2011).

The tourism industry has also continued to grow rapidly and more than 2 million foreign tourists annually visited Cambodia in 2007 and 2008 but the number declined in 2009 due economic downturn. The global financial crisis is weakening demand for Cambodian exports, and construction is declining because of a shortage of credit. The Cambodian government is working with bilateral and multilateral donors to address the country's many pressing needs. The Government's policies are directed towards strengthening macro-economic stability, promoting private sector development, sustainable development of the agriculture sector, advancing rural development, ensuring a sound natural resources management, and also encouraging income generation activities, embarking on land reform and increasing access to micro-finance for the poor. The government also promotes international and regional cooperation, especially the integration within ASEAN with partnership with China, Japan, South Korea and India, and the development of the Greater Mekong Sub-region. Rapid development and participation of the private sector play an increasingly important role in development of power sector, job creation, thus, liberating people from the shackles of poverty and improving their living standard. Furthermore, competition is the best way to avoid concentration of power, oligarchy, monopoly, corruption and other distortions. The government has formulated a policy for micro-enterprises and small and medium enterprises (SMEs).

3. Energy Sector in Cambodia

Cambodia's power supply facilities were heavily damaged by war and its rehabilitation were made under the support from the World Bank, ADB, Japan, USA and European Countries. At present, the electricity supply in Cambodia is fragmented into 24 isolated power systems centred in provincial towns and cities. All are fully reliant on diesel power stations. Per capita consumption is only about 48 kWh / year and less than 15% of households have access to electricity (urban 53.6%, rural 8.6%) and the amount of electricity consumption is as follows: Private sector 0.5%, Service sector 40%, Industrial sector 14%. The supply requirements are projected to increase in average by 12.1% per year, and the peak load is expected to reach up to 1 000 MW in 2020.

Energy sources in Cambodia can be divided into two: non-renewable and renewable energy resources. Non-renewable energy resources include fuels like LPG, gasoline, diesel and other petroleum products) and renewable energy resources are hydropower, biomass, solar, and wind. Cambodia imports fuels (e.g. LPG, gasoline, diesel and other petroleum products) in average 900 000 tons/year in 1998-2000. The country is also expecting to pump its oil by 2012 and the extraction will be carried out by Cheveron. Theoretically, the hydropower potential of Cambodia was estimated about 10 000 MW in 1995 exclude small streams and could therefore play a significant role in the long-term energy development in the context of global warming. Hydropower will contribute to reduce demand for fuel consumption which is a major source of greenhouse gases (GHG). At present only two mini-hydropower plant are in operation: O Chum II mini-hydropower plant with the installed capacity of 1000 kW has been constructed and operated since 1993. Kirirom I hydropower plant with 12 MW, was rehabilitated and operated by CETIC, a Chinese company, under Build Own Transfer (BOT) agreement for 30 years since mid-2002, together with the 120 km 115 kV transmission line to Phnom Penh. Biomass also plays an important role in generating electricity through gasification process for rural areas in the country (Abe et al, 2007). According to the measurement during 1981-88 in Phnom Penh, it showed that an average sunshine is 6-9 hours per day which indicated potential solar energy source (average of 5 kWh/m^2/day). The application of Photovoltaic system with total installed capacity of around 130 kW is a recent development in Cambodia, as donated by international organizations such as UNICEF, Red Cross, SIDA and FONDEM who installed demonstration systems on health and rehabilitation centres.

However, the potential of wind energy in Cambodia has not yet been assessed since wind energy is relatively new to the country.

Table 1: Hydropower generation candidates in Cambodia

Project	MW	GhW
Battambang I	24	36
Battambang II	36	187
Battambang III	13	77
Kamchay	180	558
Middle St.Russey Chrum	125	668
St. Atay	110	588
Lower Sesan 2	207	1065
Lower Srepok 2	222	1176
St. Chay Areng	260	13
Kirrirom III	13	70

Source: Cambodia Energy Sector Strategy (Draft)

4. Cambodia Energy Policy

The Cambodian government has been able to identify the country's major constraints to growth and development, and gather them into the Rectangular Strategy, which identifies four pillars to growth—i.e. the "growth rectangles" includes agriculture, private sector development, capacity building and human resources, and infrastructures. In addition, energy policies are recognized as one of the priorities within the latter of the pillars—infrastructure, and the government acknowledges the importance of energy in the overall development of the country. Addressing the increase of international oil prices, the government is exploring new sources of energy, including hydropower, offshore and onshore oil and gas, and renewable energy.

An effective energy policy is considered by the government to be critical to Cambodia's future economic and industrial development and to improving the standard of living of its citizens. At present, only between 5% and 10% of Cambodia's population has access to networked power and the government is strongly committed to significantly increasing the percentage of its citizens to have access to the electricity distribution network (Vichit, 2006). According to Abe et al (2007), around 76% of the 10,452 villages of Cambodia will still be without electricity in the year 2010. The government has set a target of 70% of households to access to electricity by 2030 (NIS, 1999). However, Cambodia does also need a structured and comprehensive energy strategy that can address energy conservation needs.

At regional level, the government is working closely with ASEAN and the GMS countries to connect to Tran-ASEAN gas pipeline. By doing this, the country is able to secure its natural gas supply, improve energy supply for industries and also reduce greenhouse gas. Moreover, the government is also collaborating with neighboring countries like Thailand, Vietnam and Lao to link its power grids to their hydro-electricity projects. According to Watcharejyothin and Shrestha (2009), the structure of power generation system would change from heavy dependence on oil towards the coal, hydro and biomass by 2035 and coal-based generation would account for 35% of total power generation capacity, followed by hydro (29%), oil (24%) and biomass (11%). This is in line with the current government policy in promoting alternative sources of energy for the country. The coal-base electricity project has already proposed in Sihanouk province and many hydro-power projects have been approved so far while biomass projects are being promoted in some provinces.

5. Energy Development in ASEAN

The ASEAN plays an important role in energy development of the region. The ASEAN 2020 Vision adopted in 1997 by the heads of state at the 2nd ASEAN Informal summit held in Kuala Lumpur envisioned an energy-interconnected South East Asia through the ASEAN Power Grid and the Trans-ASEAN Gas Pipeline Projects (Atchatavivan, n.d). The ASEAN Power Grid Program consists of 14 bilateral and multilateral electricity interconnection projects that cover many areas in the region. Accordingly, an integrated and coordinated planning policy approach for implementing ASEAN's Power Grid should pave the way to successfully enhance intra-regional electricity trade. It should also be useful in development of business opportunities and promotion of joint or cross-border investment in energy projects. It is also believed that ASEAN Power Grid will contribute to low down the cost of electricity in ASEAN countries as well as ensure sustainability of energy resources and contribute to energy efficiency (Yong, 2004). Meanwhile, the Trans-ASEAN Gas Pipeline Program has been integrated into the national gas pipeline. A shift from coal and petroleum will contribute to mitigation of emission problems as gas is a much clean energy source than coal and oil (Yu, 2003).

However, with the considerable concern over global warming, renewable energy sources have become more attractive for electricity generation around the globe as well as in ASEAN. Under the Kyoto Protocol, the ASEAN countries are well positioned to be benefited from Clean Development Mechanism (CDM) projects, especially in the areas of renewable energy and energy efficiency (Lidula et al., 2007). In addition, there is a need to build capacity and adapt policies that can promote these concepts in ASEAN countries. The solar photovoltaic technology is very suitable for ASEAN countries because the daily radiation of these countries in more than 4.5kWh/m² on average (Table 1). The solar photovoltaic is widely used in some countries such as Indonesia, Thailand, Philippines, and Vietnam compared to other ASEAN countries (Lidula et al., 2007).

Wind energy is also a potential alternative source to generate electricity in ASEAN and according to Cabrera and Lefevre (2002), mountain regions like central, southern and coastal areas of Vietnam, central Laos and western Thailand are the excellent areas for large scale wind energy generation.

Table2: Potential and utilization—solar photovoltaic

Country		Solar PV	
Name	Land area (km²)	Potential solar radiation (kWh/m²/day)	Utilization
Brunei	5765	No information	Hybrid system (2.4 kW solar & 80 kVA diesel)
Cambodia	181,035	5 (6–9 h)	700 kW solar PV
Indonesia	1,890,000	4.8	5 MW
Lao PDR	236,800	4.5–4.7	285 kW solar PV
Malaysia	330,257	4.5	1.5 MWp (PV stand alone) 450 kWp (grid-connected PV)
Philippines	298,170	5.1	1 MW (centralized solar PV)
Singapore	697.1	No information	90 kWp
Thailand	513,254	5.1	6 MW
Vietnam	330,363	5 (4–5.9 h)	0.6 MW (solar PV)

Source: Lidula et al., 2007

6. Major Challenges in Promoting Energy Conservation Policy in ASEAN

The main challenges for energy conservation policy in ASEAN include lack of funding, lack of experience and knowledge and limited policy frameworks are the most common (Lidula et al., 2007). In addition, given that the wide gap in economic growth, renewable energy technology may be appropriate for some countries while others are not able to afford the cost of investment in the technology. Table 2 summarizes the challenges of ASEAN in promoting renewable energy and it indicates clearly that some of the barriers are common to most of the countries in the ASEAN. Furthermore, most of these issues can be addressed by improvement in policies and regulations in renewable energy and energy efficiency. Supportive policies and regulations at country level and regional level also play a key role in promoting renewable energy (e.g. Thailand and Malaysia).

Table 3: Major barriers in promoting renewable energy in ASEAN

Description		Lao PDR	Vietnam	Cambodia	Malaysia	Indonesia	Thailand	Philippines	Brunei	Singapore
1	Lack of experience and awareness in technology and management	✓	✓	✓			✓	✓	✓	
2	Lack of funding or financing difficulties	✓	✓	✓	✓	✓	✓	✓		
3	Limited policy framework		✓	✓	✓		✓	✓	✓	
4	Lack of institutional, financial and technical structures to promote RE	✓	✓		✓		✓	✓		
5	Reliance on national grid and lack of private sector participation (no market infrastructure)		✓			✓	✓	✓	✓	
6	Inadequate data and information	✓	✓	✓					✓	
7	Reluctance to invest because of high investment cost	✓				✓	✓	✓		
8	Low efficiency or quality of some RE		✓				✓	✓	✓	
9	Insignificant utilization of RE (not financially sustainable)		✓	✓				✓		
10	Lack of research personal or trained man power and lack of R&D in some RE	✓	✓				✓			
11	Fossil fuel subsidies	✓				✓		✓		
12	Taxes on imported equipment or no waiver on taxes		✓					✓		
13	Inappropriate distribution facilities	✓								
14	Political involvement in reform agenda							✓		
15	Legislation issues in connecting RE to National grid				✓					
16	Objections from the public to have power plants in the area						✓			
17	Lack of government support						✓			
18	No economically viable RE									✓
19	High total installed capacity resulting no requirement of new power plants									✓

Source: Lidula et al., 2007

7. Conclusion and Recommendation

The increasing demand for energy will trigger the increase of CO_2 emission needs in the region. As a result, ASEAN countries should involve energy generation project with CDM which is not only benefit energy conservation but also mitigate climate change. Furthermore, diversification of energy sources, particularly from renewable sources will also reduce the demand on fuels and promotion of clean coal technology among ASEAN countries should also be taken into account. Even though ASEAN countries are enrich with renewable energy resources, these resources are underutilized by some countries. As a matter of fact, exchange in renewable energy technology should be made among the members to fully use renewable energy resources effectively.

Reference

Abe et al. (2007). Potential for rural electrification based on biomass gasification in Cambodia. *Biomass and Bioenergy* 31 (2007) 656–664

Atchatavivan.P.(n.d). ASEAN Energy Cooperation: An Opportunity for Regional Sustainable Energy Development. Available at: http://www.pon.org/downloads/ien15.8.Atchatavivan.pdf

Cabrera M.I. and Lefevre T. (2002). Overview of wind energy in Southeast Asia. Available at http://www.ec-asean-greenippnet.work.net/dsp_page.cfm?view=page&select=126S

Cambodia Daily. (2011). Rubber Exports Rose 35% to $73 Million in 2010. January 6, 2011, Pages 29.

Cambodia Energy Sector Strategy (Draft) (n.d.). Available at: http://www.un.org/esa/agenda21/natlinfo/countr/cambodia/energy.pdf

CIA .(2010).The World Factbook. Available at https://www.cia.gov/library/publications/the-world-factbook/geos/cb.html

Lidula et al. (2007). ASEAN towards clean and sustainable energy: Potentials, utilization and barriers. *Renewable Energy* 32 (2007) 1441–1452.

National Institute of Statistics (NIS).(1999). General population census of Cambodia. Phnom Penh, Cambodia, National Institute of Statistics (NIS) of Cambodia. p. 367.

Vichit, H. (2006). Oil and Gas Prospects in Cambodia: Overview of status of current exploration and future plans for oil and gas development and production. ASEAN Energy Business Forum 2006: Meetings- Conference – Exhibition, 26-28 July 2006, Vientiane, LAO PDR

Yong.O. K. (2004). Speech on "Integrating Southeast Asian Economies: Challenges for ASEAN". Remarks at the ASEAN Energy Business Forum 2004. June. Manila, Philippines. http://www.aseansec.org/16139.htm

Yu.X. (2003).Regional cooperation and energy development in the Greater Mekong Sub-region. *Energy Policy* 31 (2003) 1221–1234

Watcharejyothin. M and Shrestha.R.M.(2009). Regional energy resource development and energy security under CO2 emission constraint in the greater Mekong sub-region countries (GMS). *Energy Policy* 37 (2009) 4428–4441

World Bank. (2010). World Development Indicators 2010. Washington DC: World Bank.